MAY ROBOTS HELP HUMAN TO BRING ECONOMY AND

ABILITY IMPROVEMENT

JOHN LOK

Made with ♥ on the Notion Press Platform
www.notionpress.com

Contents

Preface

What methods may help human to create clever as well as creative ability? I shall indicate robot technology invention how to train human creative ability , also I shall explain whether robot tool may help human to improve economy development or help businessmen to reduce cost. Readers can make analysis whether robots may help human to bring long term economic benefits and improve ourselves ability and reduce cost for businesses.

Prologue

Content of content

CHAPTER ONE

Introduction brain imagination creative effort

What does brain imagination art creative ability mean ?

Can we train ourselves brains to raise more art creative abilities? One excellent author, painter, music writer, clothing designer, house designer, dance performer etc. different kinds of art creative performers, whether they use some methods or skills to raise their brain imagination art creative effort or it is themselves owning genius brain imagination art creative abilities from their born or birth date.

It is one interesting question concerns how to raise ourselves brain imagination art creative efforts. Firstly, we need to know whether what brain imagination art creative effort means. In fact, many people feel some art genius, their imagination art creative abilities are due to their parents give them. It means that when they born, they must be genius, any kinds of art imagination creative efforts that they must own, e.g. some genius own writing story creative content ability, creating song or music ability, creative beautiful paints ability, designing clothes or houses or any things ability, creative dance performance ability.

However, some brain scientists or brain doctors or psychologists explain that evidences and experiments indicate may of art geniuses, their art imagination creative abilities are due to their learning more than themselves born to own from parents.

What does brain imagination creative effort mean? Brain imaginations art creative abilities may include many kinds of brain creating imaginations. For musical imaginations example, it may explain any creative aspects of music listening in the activities of composition, improvisation, and performance. So, the music or song writer can own good brain of music creative effort to write many different nature of musical beauty to bring enjoying and listening song or music emotion to music or song listeners.

So, brain imagination may include music or song creativity to any one. But, the differences between good music or bad music creativity may due to the music or song creativity may due to the music or song imagination creator whether who owns what level of musical knowledge, training, literacy, writing music or song experience, or playing music performance experience to the music listener individual listening music or song taste etc. different factors to bring individual emotion response to feel whether the song or music is good or bad after the music writer finished to write the song.

So, any kinds of song or music creativities have close relationship to the creator's brain music or song imagination creating effort had how much. It means that of the music creator owns high level of brain music imagination creative effort, then the music creator ought have enough effort to create many good song or good music to let any one listener to listen and feel their song or music creativities can own unique listening feeling to compare general music creators their common music or song creativities.

On theory explanation, brain imagination may mean that creative thinking, creating thinking is defined as the competence to engage any kinds of brain imagination productively in the generation, evaluation, and improvement of anyone of brain imagination, ideas, that can result in original and effective solutions, advances in any kinds of knowledge imagination , e.g. design a house, design a cloth, design a product, writing one story, writing one music or song, design a dance performance, painting a picture etc. different kinds of imagination.

Imagination may be listening imagination, e.g. song, music or reading a story book, wearing a dressing cloth, living a house, seeing a dance performance. Hence, imagination may be touched or seen or felt by any one. Hence, any one creative art performer must need have excellent creative thinking to create their product imagination, e.g. how to write one good story, how to write a good listening song, how to design a house or a cloth or a product, how to prepare a dancer performance etc. different kinds of creative product imagination to let any one customer to feel their creative products can have the unique or excellent quality to compare other general similar or same kinds of creative products.

Thus, imagination is the seed of creativity. Indeed, there is one fundamental skill that makes creativity possible. Without imagination, there can be no creativity. Imagination refers broadly to the human capacity to construct a mental representation of the which is nor currently present to the senses (Markman, Klein, & Suhr, 2009; Seligman et al., 2016).

Across social –emotional domains, there are a number of forms of imaginative thought, include thinking informed by an understanding of multiple cultures, pretend play, prospection, memory construction, counterfactual thinking, and mind wandering (Abraham, 2016 ; Runce & Pina, 2013). Many forms of imagination, specially imagination about people, including oneself, across time and space, draw heavily on the brain's default mode network, a network composed of several brain regions along the midline of the medial prefrontal cortex, medial pertietal cortex (Andrews, Hanna, Smallwood & Spreng, 2014; Zmmordion-Yang Christodoulour, & Singh 2012, Raichle & Snyder, 2007, Schactot, Addis & Buckner, 2007). Other forms of imagination that involve visualizing, physical objects or physical space are thought to recruit more heavily the brain's executive attention network and dorsal attention network, a network involving communication between the frontal eye fields and the intreperietal succs (Andres- Hanna et al., 2014: Jack et al 2013).

Hence, many brain doctors and brain scientists and psychologists imply that imagination is the seed of creativity. Genius's unique creativities can be caused by their brain imagination. Any one creative product creator whose brain may be trained to own unique creative effort by themselves learning or nay new creative knowledge skills or methods in order to achieve to raise themselves brain imagination creative efforts.

On conclusion, it seems that why any one creative effort is due to creator whose imagination creative effort is due to they attempt to learn new creative knowledge more than their born genius to own creative effort. Many brain doctors or brain scientists or psychologists began to research how to raise ourselves brains to achieve owning excellent imagination creative effort. I shall attempt to explain how we can attempt to learn in order to raise ourselves brains imagination creative efforts in order to become one excellent painter, music/song writer, author, designer, dance performer etc. different kinds of creative occupation performers.

CHAPTER TWO

Training brains creative imagination effort methods

Can we apply some methods or skills to help ourselves brains to raise imagination creative effort? In fact, many brain doctors or brain scientists or psychologists began to research whether we can apply what skills to improve ourselves brains imagination creative efforts. They had began to attempt to find any one to attempt to do any brain imagination improvement experiments . Their aims to find the best skills to help human to improve brain imagination memory in order to create high level art performers. If one day, human individual brain can be trained to raise brain imagination memory function effort, then many general level of authors, painters, music or song writers , art performers, designers , their skills can be improved to be one proficient art work creator easily. Because if they can be confirmed that one kind skill or some skills can help themselves brains to raise imagination creating effort significantly. Then, when every common art work creator occupation performer can learn any useful skills to attempt to raise their brain imagination creative effort, it can bring much benefits to our societies because when many common art creative performers can confirm to improve their brain imagination creative effort from some skillful trainings. Consequently, we will have much beautiful cloths to wear, good design of houses to live , good music or song to listen, see beautiful drawing pictures, good books to read , good dance performances to see, because all of these creative art performers, their imagination creating effort can be improved significantly. Then , our social cultural level will be influence to raise in long time significantly.

The question concerns whether what skills or methods may help ourselves brains to raise imagination creative efforts? I shall attempt to indicate some possible skills or methods, they may help ourselves brains to raise imagination creative efforts as below:

For dancer performance behavior example, it is at its simplest, allowing your child to thrive though constant play and exploration, uninhibited by the strict rules of science, reason, law or a judgmental society. For any one dancer, it requires an open channel to allow dancer behavioral information and ideas to flow freely into the dancer's space. It requires an open mind and a sense of freedom and limitlessness. So, one proficient dancer needs to own high dance skillful memory to remember every whose dance behavior as well as learn how to create good dance skills in order to attract audiences attention and satisfy their visual dance enjoyment feeling on the theatre hall.

Hence, dancer needs to raise whose dance skillful imagination creative effort, it means that they need to learn how to excite themselves brains to create dance behavioral imagination in order to improve themselves dance skills effectively.

I believe that strongly in the idea of every human action stemming from a complex environmental factors. I also believe an extra ingredient is somehow part of imagination. An internal imaginal word, so such as one proficient dancer case, if she can train herself brain to remember every high level dance steps easily. Then, she ought improve herself dance skills in short time rapidly. So, training to remember dance steps , which is needed to help the common dancer to become one proficient dancer more easily. For dancer case, she may use imagination to recall emotions needed in interpreting a character for any one dance performance to picture an overall aesthetic before it comes to fruition, to invent new artistic dance concepts, to invent new movements to connect old movements , to picture the dancer whose body executing a movement before she has even done the dance steps, to try something new dance steps. Hence, new dance steps and dance bogy behaviors from whose old dance st4eps and body dancing behaviors in order to achieve to feel her dance steps and body dancing behaviors had been improved significantly. It is one good skill to improve anyo0ne common dancer dancing skills, when she can train herself brain to create good dancing steps and dancing body behaviors as well as remember every old dancing body behavior or dancing steps in order to raise how to improve or create new dancing body behavior or new dancing steps. Consequently, she can learn how to change new dancing body behaviors in order to improve herself old dancing skill easily. Hence, any one dancer must need to raise whose brain dance imagination creative effort in order to improve whose old dance skills significantly.

Research concerning the role of memory in imagination or brain imagination creative effort skill issue, it is future interesting brain behavioral research issue for any one brain doctor or brain scientist or psychologist. They aim to help our societies to produce more excellent art creative performers. Most psychological theories of imagination can be seen as theories of imagination consider that " creative" imagination, e.g. paint, writing a story, playing music, designing a house, designing a cloth, designing a dancing steps. All of any these, they belong to " creative imagination". Their imagination is needed to create by ourselves brains. When ourselves brains have clear picture, then the author can follow his brain picture or mind to create a story content, write a song, design a house or a cloth or organize a dancing steps.

Hence, " brain imagination picture" may be one main element to help any one creative performer to create any kinds of creative product very easily. If the creative performer can own good brain picture memory, then his story content can be more attraction, his dancing steps can be more attraction, his house or cloth or any kinds of products design can be more attraction. The question concerns how to create attractive brain picture imagination? Hence, when the art creative performer can have good brain imagination, then he/she ought have good creativity or creative effort in order to create whose new story , new song, new design house of cloth or new dancing steps very easily.

`In fact, imagination pervades human experience. Children begin engaging in pretend play and although cultural and parental attitudes affect the amount and content of imagination play. As adults, we are consumers and creators of fiction, song, story, house, designer, dance performance etc. and we respond emotionally to imagined scenarios. Moreover, we invent fictions even in the pursuit of facts, face of neurological disorders, in defending the bases of our decisions, and in the construction of autobiographical memory.

However, by applying useful knowledge in extraordinary events with heightened emotional content, learners may be better able to access important cultural skills or facts. Consistent with this, researchers have suggested that imaginative engagement might support a range of cognitive abilities, including creativity, intelligences, problem solving, symbolic reasoning, language development , theory of mind, narrative skills, social skills, causal reasoning , emotional regulation , and executive function.

I believe that thinking of new ideas is not an optional exercise in creativity. It is fundamental to learning . The learner must need gather new information in order to raise whose brain imagination creative effort. So, gathering new information is the best method to help our brains to raise imagination creative effort. It is more effective to compare how to think of new ideas to achieve how to create ourselves brain imagination creative effort. For a dancer, if she can attempt to gather any new dancing styles new information from internet channel daily. I believe that she can improve whose dancing steps in order to create new dance style more easily. Otherwise, if she only concentrates on how to think her new dance style ideas by herself mind. I believe that she can not create any new dance attractive style to improve her old poor dancing style easily. Hence, gathering new information may be another useful brain imagination creative effort improvement skill.

Another kind of method is that teaching anyone to new skill. One of the best ways to expand your learning is to teach a skill to another person. After you learn a new skill, you need to practice it. Teaching a new skill to others needs you to explain the concept and correct any mistakes you make. This can improve your mental activeness to a great extent. Next method is that spending spare time for physical activity, many studies have confirmed that daily physical activity also keeps the brain sharp and active. The simple science behi8nd this is that physical activity accelerates the circulation of oxygen on the mind. All of these methods can help your brains to improve memory and raise brain imaginatio0n creative effort effectively. Then, we can improve memory, it can assist us to raise creative thinking or creativity.

Creativity means the ability to change traditional ideas, rules, patterns, relationships and to create meaningful new ideas, forms, methods etc. originality, progressiveness, or imagination. In fact, every one has the capacity to be creative. In the psychological view, these is a debate over whether anyone is born with innate creativity or if everyone who has it has developed a talent. Though some scientists believe certain individuals have a higher aptitude for creativity, many attest that creativity is an actually a skill and anyone can learn a skill. This is the integration of memory and creative thinking at work.

Consequently, we tend to think a parts of the brain specialized for one thing, one particular function at a time. However, neuroscentists attest that all parts of the brain are constantly interacting and building strong neural pathways is the best way to keep all parts of the brain healthy. Finally, I shall indicate ways to exercise creative thinking skills, e.g. changing your routine, you will need to experience anything new or give yourself a new experience to active your brain in creative process. Try changing something small or adding a new activity each day. Thus will help your brain and your body to stretch your creative muscles, or read a book or listen music, anyone of these daily habit behaviors, they will help your brains to raise imagination creative effort effectively. Hence, we can not neglect any one of these simple skills or living habits, they may help ourselves brains to raise imagination creative effort indeed.

CHAPTER THREE

Can exercise bring more creative effort to brains

The relationship between enough sleeping and
brain imagination art creative ability raising level

Can keeping enough sleeping time raise brain creative imagination effort or raise memory? Can enough sleeping influence our brains own more art creative ability, e.g. creating good story content ability, drawing beautiful paint image ability, creat5ing good song or music ability, creating beautiful house or cloth or any kinds of products ability? Many brain scientists had begun to research how enough sleeping time has direct or indirect relationship to improve ourselves brains creative efforts. They also believe that it is possible that enough sleeping time may help ourselves brains to improve creative e and memory effort. If it is true, whether we need to sleep how many hours in order to improve ourselves brains creative efforts. Otherwise, if we lack enough sleeping time, our brains creative e efforts will be influenced to poor? For example, one owning many years writing experience proficient author, if he can not have enough sleeping time every day, it can bring poor creative story imagination mind ability to recreate any new story content in possible. It is one interesting question concerns the relationship between enough sleeping time and brain creative ability research? I shall attempt to indicate evidences to explain their relationship as below:

` Any one must need sleep. If one lacks at least 8 hours sleeping time in the day, he won't have enough nervous to do any matters, even jobs. So , it seems that any creative tasks, e.g. writing stories, painting pictures, designing houses, cloths etc. creative jobs. The art creator must need have enough time to sleep in order to prepare to raise his brain imagination creative feeling to achieve how to design one beautiful house in order to attract people to choose this house to buy in preference, or design one beautiful cloth to attract people to choose his design cloth to wear in preference, how to write horror story content to let readers to feel fear or write romance story content to let readers to feel they are lovers both or write space scientific story content to let readers to feel that they are catching rocket to fly to outer strange space environment to carry on one time space existing journey, even how to design one time attractive dance performance to let audiences to feel all dancers to feel all dancers are performing attractive dance steps or create soft music to let music listeners to feel this music is soft or comfortable to listen, when they are sitting in the theatre to listen this music performance. So, all of these creative tasks to the art creator, who must need have clear brain imagination or nervous to help them to create any one of those creative product in order to let customers or audiences feel their creative products are beautiful or attractive to compare other some creators whose art products.

Hence, art creative imagination must need to any one art creator, if he/she hopes that his/her art creative product can bring more attraction to any one art product buyers or audience to consider more to compare his/her other same art creative product competitors. So, any one art creator must not need to own high educational level of working experiences . Otherwise, they must need top own high level art imagination ability to compare general people. Hence, it is true that any one art creator must need have more clear and good imagination effort to compare other occupation working people in our societies, if they hope to own good creative imagination effort to attempt to create their any kinds of creative products, e.g. design of a house, a cloth, write one story etc. creative products. So, it is ensure that creative art product ought need have good brain imaginatioOn to compare general occupations.

In fact, many brain scientists or brain doctors or psychologists believe that enough sleeping time, it can influence any one whose nervous next day, such as they had attempted to carry on many brain imagination creative effort researches or experiments . They confirm that if the art product creator or art entertainer can have enough sleeping time, it can help the art product creator or art entertainer to create good dance performance, good listening soft music, one beautiful design house or cloth , dancing a good dance performance to compare lacking enough sleeping time product creator or art entertainer. Hence, it seems that enough sleeping time may assist any one art creator to raise brain imagination creative effort.

What the methods to train brain creative effort?

How we can train ourselves brains to raise high creative efforts? Can we train ourselves brains creative efforts by learning method? Has it close relationshipship between ourselves bpdies and ourselves brain creative efforts? Do ourselves brains creative efforts to be poor if we have no health bodies?

I believe that we must need have health bodies, then our health bodies may bring more creative effort to ourselves brains easily, because health bodies may help us to raise ,memory ability, keeping happy and pleasure positive emotin to do any things every day. Due to our brains must be our part of bodies. Our brains are inside to our bodies. Although, we can not see our brains , but we can feel oursleves brains are working, if our memory is high, e.g. we can remember our teachers what they had taught all contents after every lesson. Then, our examination results must be improved because we have good memory . So, enough sleeping time is one good method to raise ourselves brains' memories.

In fact, enough sleeping time does not needed to have enough training. So, training our brains to raise creative effort method, we only need have enough sleeping time in order to keep our brains have enough nervous to remember any thibgs more easily, e.g. studying is one good example for remembering training action or behavior, we can attempt to train ourselves brains to remember any new knowledge from teacher individual teaching. When we learn any new knowledge in lessons, we are using ourselves brains to attrmpt to remember what the teacher is teaching in lesson . If we brains feel tried, we can have breaking time for less sleeping time between lessons in order to keep ourselves brain memory longer time.

In fact, ourselves brains memory and brains creative effort whether they are high or low level, they have close relationship because our creative effort may be influenced to raise if our memory effort can increase rapidly and it can be kept longer time. Many brain doctors, psychologists, brain scientists had researched to attempt to do experiments to confirm that brain creative effort and brain memory effort have close relationship. Hence, many of them begin to believe that if one person can have high memory effort, then whose brain ought have hifh creative effort, e.g. creative story writing ability, creative drawing paint ability, creative music or song ability, creative designing house, or cloth or any products abiliity . All of these different kinds of brain creative abilities , they must have absolute relationship to brain memory. It means that if your brain can have good memory ability to remember or learn any kinds of new knowledge rapidly in long time, then your brain can have high creative ability to be trained to create any kinds of new knowledge easily, e.g. writing fun story, writing fun or good listening song or music, drawing beautiful paint, design attraction of house or cloth etc. any products, even organizing good dance performance. All of these creative tasks must need have good memory ability to the creator, hence, enough sleeping time must be the best method to help us to improve oursleves brain memory or brain creative effort both. We can not neglect to keep enough sleeping time in order to raise oursleves brain memory and creative effort both, if we hope to become an excellent story writer, music or song writer, house or cloth design, good organized performance dancer.

On conclusion, attempting to apply brain memory effort to do creative effort skillful tasks, when one person can have enough sleeping time every day, then his brain memory ability may be influenced to raise or improve. He can attempt to do some simple creative task, e.g. learning how to draw one natural environment scene paint, learning hoe to create a horror story, learning how to create a good listening song or music, even learning how to organize every steps for the dance performance etc. different kinds of creative tasks behaviors. Because when the person feels that he has enough sleeping time, his memory ability may be influenced to increase memory ability. When one house designer, he feels that his memory ability is increasing, he ought remember all prior beautiful or urgue house design in his memory if he is one house designer. After this house designer's prior all house design drawing picture will be remembered again. So, this house designer can remember all his poor drawing paints in order to improve or understand whether how he ought continue to draw or improve his new house picture in order to design his new house design more easily. Because when one house designe needs to spend time to create or design one new house. If he can remember which kinds of house , he had drawn in past in his brain memory. Then, he can avoid to repeat to design his past old house design again. He can use his new house design method to create another new house paint design. So, this house designer can improve his new house paint design if he can remember what kinds of houses, he had painted to design in his past. Then, he can avodi to repeat to design the kind of similar or same style of house again. Consequently, one good brain memory house designer can avoid to design same or similar house again, then this house deisgn may have more house design creative effort in order to design his next new house easily. So, it explains why it may have close relationship between brain memory and brain creative effort.

Can exercise bring more creative effort to brains

Any excellent occupation people who must need to often do exercises in order to achieve excellent performance, such as sportman, doctor, lawyer, accountant, engineer, teacher, architect, singer etc. different professional occupations. All of these occupation people, who must need to spend time to do exercises repeatly again in order to achieve proficent performance or proficent skills. The question concerns whether of one creative art working person who often do exercise, then he/she can bring more creative effort to themselves brains, e.g. one authoer often does imagination to feel new things in order to attempt to write any new story. In his/she creative imagination process, whether himself/herself brain can be influenced to raise or improve or increase effort when he/she often uses his/her brain to mind any new science story scene to achieve brain creative effort or if one musican often uses himself/herself brain to attempt to mind any new music suddenly. When he/she feels himself/herself brain has one good music or song suddenly, then he/she writes down the music or song in order to avoid that he/she forgets when himself/herself brian has this new music or song imagination in his/her mind. So, he / she often does music /song creative exercise in order to improve his/her any new music/song imagination. Can this musician often do music/song creative imagination to bring high brain creative level? For a house deisgner example, if this house deisgner often does brain creative exercise to draw any house design or house picture when he /she feels himself/herself brain has any unique house imagination suddenly. Can his/her frequent drawing any new house picture design behavior, which changes hisself/herself brain creative effort to be raised significantly? So, such as my explanation that it is possible that any exploration that it is possible that any person can attempt to do brain creative exercise in order to raiee our brain creative effort more significantly and easily.

Why do frequent brain creative imagination exercise, they may help oursleves brains to raise brain creative effort more easily or effectively? I shall attempt to explain as below:

Our brains are similar to our bodies, e.g. hand, foot. We must need often do exercises, in order to let we can run rapidly or climb mountains safely or we can swim rapidly. So, if one sportman hopes to win the sport competition, he/she will often run or swim or climb mountains or ride bicycles before this sport competition will begin. So, our brains are such as our part of bodies. Ourselves brains must often need to be done exercises in order to raise creative effort, if we hope tobecome one proficient creative author, drawing painter, house designer, cloth designer , musician etc. art creators.

Our brains must be needed to do any creative mind exercises again and again every day in order to improve oursleves brains imagination creative effort significantly. Any one proficient art creator, who must not be talent especially, they must not own unique talent characteristics. Otherwise, some of talent art creator , e.g. author, musician, house or cloth designer, painter et.c they may be foolish people in general. But, if they can often keepto do brain creative exercises, when they feel any new things in themslves mind suddenly, e.g. one special house picture imagination, one special beautiful cloth picture imagination, one good listening song or music imagination or mind. They can write down to record on paper immediately. If they can keep to write on paper to record any new picture imagination in habit. Their brain creative imagination exercise behaviors, which can help them to train their brains to learn how to create new imagination picture skills, e.g. let them to feel to create any new music or song easily, let them to feel to draw or design any house or cloth imagination easily, let them to feel to create any new story content easily, let them to feel to organize one unique dance performance easily.

What factor may cause any one of above art creator to feel how to create any new creative product easily? The main factor is that " exercise" , due to they can often train their brains to learn how to attempt to create new story content, new music or song , new house or cloth design, new dance performance stepping. When they often use themselves brains to create any new creative products, and write down on paper or draw on paper in order to record their any new creative products as well as aovid to forget their any new creative products. Then, they can revise their past creative products in order to improve their past creative products to be better.

In their " brain creative process exercises", they must help any one of these art creators to raise themselves brain creative effort significantly. Hence, " brain art creative exercise" may be one good method to help any one art creator to raise their art creative skills.

Psychological methods raise brain art imagination creative effort

Can habit leisure hobbies bring more brain imagination effort

How to learn effort in order to raise more brain imagination? Can learning bring more brain imagination effort? For example, if one person often needs any another person writing stories, in his/her reading process can it raise his/her writing ability? If one person often listens music or song , in his/her listening music or song process, can it raise his/her writing music or song creative ability? If one person often watching horror movies, in his/her watching horroe movies pricess, can it raise his/her creating horror movie imagination effort? If one person often sees cloth magazine photos, in his /her seeing cloth photos process, can it raise his/ her design cloth creative effort? If one person often sees house magazine photos, can it raise his/her design house effort? If one person often sees dance performance, in his/her seeing dance performance process, can it raise his/her organise dance performance effort? If one person often sees paints, in his /her seeing paint process, can it raise his/her drawing paint brain art creative process?

Some brain doctors, brain scientists, psychologists begain to research whether if general people often to read books, listen music/songs, see movies, see house or cloth magazines, see dance performance, see paints etc. different hobbies, when whose these different kinds of hobbies become leisure behaviors, whether their these different kinds of leisure behaviors can influence themselves brains raise imagination creative efforts. Some brain doctors or brain scientists had confirmed that when general people can spend about 2 to 3 hours per day to read stories, listen music/song, see movie, read cloth or house photo magazine, see paints, see dance performances etc. different kinds of hobbies or lesiure behaviors. Then, these general people whose brains can be influenced to raise more imagaination creative efforts, .e.g. creating story content effort, creating music or song effort, designing house or cloth effort, organizing dance performance effort, creating movie content effort, creating paint picture effort, because when general individual can spend time to do any one of these leisure actions to be habit bobbies every day. Although , they do not be trained or taught from teachers, or they do not born to own high creative effort, but when general individual can attempt to spend time to feel enjoyable to choose to do any one of these leisure action to become hobbies. Then, their brains may be influences to raise writing story, writing music/song, drawing paint, organizing dance performance, design house or cloth etc. different aspects of art creative efforts.

On conclusion, when one general individual can often spend time to do leisure actions or behaviors to be hobbies, then they may help themselves to raise brain imagination creative effort in possible. So, we can attempt to choose any kinds of leisure interest to be leisure hobbies as well as do leisure actions in habit, then our leisure habits will help oursleves brains to raise art creative efforts in possible.

Training and teaching methods raise brain imagination creative effort in possibility

Can general people be trained or be taught in order to help ourselves brains to raise imagination creative efforts? Many brain scientists and brain doctors had began to find general people to do experiments concern whether any general people whose brain imagination effort can be influenced to raise after they are taught or are trained, e.g. learning drawing paint skills, learning dance skills, learning creative story content skills, learning house or cloth or product design skills etc. different kinds of imagination behaviors. However, they discovered that although any one can learn general writing story, writing music or song , design house, cloth or any kinds of products skills, draw paint, or organize dance performance etc. art creative skills, but it does not represent that their brains can be influenced to raise imagination creative effort. The reason is that it is different between raising creative brain effort and raising creative skills effort.

In general, we can learn general writing book, writing music or song, designing house or cloth or any kinds of products, drawing paints, dancing etc. art creative skills or art creative methos by teaching or training methods. So, art teacher may let us to learn any kinds of art creative skills, but, if we hope to raise or improve ourselves brain art creative effort. It is not possible that teaching and training methods both may hepp us to raise brain art creative effort, because human brain is one part of ourselves bodies. Brain is not hand or foot, we can use hand and foot to attempt to do any art creative tasks, e.g. how to use hand to draw one beautiful paint, hoe to use foot to dance, how to use hand to design one beautiful house or cloth or any kinds of products. So, learning art creative skills, it means that learning how to use hand or foot to do any art creative behaviors. We can only learn art creative skills hoe to use hand to draw beautiful paint, how to use foot to dance, how to hand to design one house, one cloth or any kinds of products. So, teaching and training methods can only let us to learn how to use hand or foot to do art creative skillful behavior more easily. These both methods can not help ourselves brains to raise brain art imagination creative effort in possible.

Nowadays, many brain scientists and brain doctors began to believe that teaching and training both methods only help oursleves to learn how to use hands to improve drawing paints behaviors in order to achieve more easily or how to use foots to improve dancing behaviors in order to achieve more attractive dance performance or how to design cloth or house or any kinds of products more attraction, or how to use our mouths to sing more attraction of sonds. All of above these art creative behaviors are only " art creative skills". When we are taught or trained to learn any one of these art creative skills by teachers, ourselves brains won's be influenced to raise any kinds of art creative efforts. So, teaching and training both methods can not raise ourselves brains creative effort. Otherwise, these both methods may only help ourselves hands and foots to improve art creative skillful behaviors in order to achieve how to use hands to draw paints more easily, how to write story content more attraction, how ro use hands to design house or cloth or any kinds of products more attraction, how to use our mouths to sing songs more clearly. If we hope to improve or raise ourselves brains art creative efforts to be any kinds of art creator, we can only keep enough sleeping time every day, keeping habits to read books, keeping habits to see any cloth or house magazines, keeping habits to listen any kinds of music, or see any dance performance etc. different kinds of art leisure activities, because when we can enjoy to do any kinds of art creative leisures when we spend nervous to do these leisure activities in habit. Then, ourselves brains can be excited to raise any kinds of art excited to raise any kinds of arr imagination creative efforts by other artists. Consequently, ourselves brains art creative minds may be influenced to improve , even raise brain art creative effort , due to ourselves brains had saved more different artists whose art imagination of memories.

CHAPTER FOUR

Brain creative ability science development

Improvement IQ raises brain art imagination creative effort development

Can we improve ourselves IQ in order to raise ourselves brain imagination creative effort? Do IQ and brain imagination, chich has cause and effect close relationship?

Nowadays, many brain scientists and brain doctors began to attempt to find young people to do experiments to research whether their brain art imagination creative effort cab ne raised if their IQ can be trained to improve their mind analysis and logic judgement effort. In results, they conclude unique conclusions, their investigations discover that if one young person whose IQ can be trained to climb up to above 100 marks, then their mind and analysis and logic judgement effort may be influenced to raise, even their brain art imagination creative effort may also be influenced to improve in possible. Hence, many of brain doctors or brain scientists or psychologists begin to believe that training on IQ improvement method can bring positive art imagination creative effort influences to develop our brain's mind analysis, ligic, judgement efforts. So, if one person , he/she can be trained to improve his/her IQ mind effort from his/her child age stage till to adult age stage. Consequently, his/her brain art imagination creative effort may be raised to more 50% of his/her general brain art imagination creative effort level.

It brings this question: WHy can improve brain IQ level to influence oursleves grain art imagination creative effort to be raised? The answer is simple that I assume the one young owns
high IQ level, his/her mind analysis and logic judgement effort must be better than general low IQ level people. SO, when he/she owns high level of mind analysis and logic judgement effort. He/she can spend less time to make more accurate judgement to solve any challenges, because his/her mind analysis and logic judgement effort compares to general low IQ level people is higher. So, when the young can be trained to reach high IQ level intelligent young, he/she can make much accurate mind analysis, logic judgement effort to know how to draw paints or pictures to let many audiences feel more beautiful to his/her creative pictures, how to create horror story content to feel readers to feel, how to sing or create the song in order to let listeners to feel
enjoyable to listen the song, how to organize the dance steps in order to attract audiences to see the dance peformance, how to design the house in order to let the house buyers feel the house design is attraction, how to design the cloth in order to let the cloth buyers feel comfortable to wear the cloth etc. different art creative tasks. So, it seems that it has cause and effect relationhsip between'IQ
improvement and brain art creative effort improvement . Because when we born, we must not own high IQ level or high brain art creative effort level. We must need to spend long time
to learn or to be trained to improve ourselves brain IQ level or brain art creative effort. Due to ourselves brain IQ level and brain art creative effort level will not be brought when we born in first day. I mean that we need time to be taught or be trained in order to improve ourselves brains IQ level or brain art creative effort level. Hence, IQ and art creative efforts have similar characteristics , such as intangible, feeling. So, we must need oursleves brains can be trained to bring high IQ level in order to improve oursleves brains art creative effort in possible. ON conclusion, improvement IQ level

can be one kind of good method to help ourselves brains to improve art creative efforts significantly.

Creative drawing paint effort brain development

A painter hopes to learn how to use hands to draw any kinds of beautiful paints. INstead of colour choice, paint pen choice, painting drawing quality of paper choice, painting tools factor.
Can the painter himself/herself brain mind creative effort influence his/her paints pictures which they can attract audiences to feel himself/herself paints are beautiful to compare general painters. IN fact, drawing paint quality of papers, painting pens, colour quality of these painting tools factors may influence the picture is beautiful or attractive or not, but we can not
negligent that the painters whose brain mind creative effort also may influence their any one paint creative feeling. For exmaple, if the painter's brain can have good creative house effort, then his/her any house design will attract more people like to see his/her any house design picture to compare other house designers. Hence, this painter ought concentrate on attempting to design any kinds
of house pictures order to improve his/her any kinds of house design picture, due to his/her brain owns high mind creative effort to create any kinds of house to draw more easily to compare
other kinds of things. So, if he/she forgets to continue to draw any kinds of house pictures, then he/she chooses to draw natural scence or human face or human body or any things, such as cloth,furniture etc. paint pictures. I believe that this painter can not draw these kinds of pcitures to compare drawing house pictures more beautiful or attraction. Hence, it seems that an yone painter needs to know that whether the painter , his/her ability can draw which kinds of paint pictures more excellent or more proficient in order to concentrate on improving to draw this kind of paint pictures. Then, it brings this question: How doe this painter know that what kinds of painting pictures who can draw more proficient? The answer ensures be "what kinds of picture image, he/she can

own more mind creative effort. Hence, the painter needs to know that what kinds of brain image that he can own more creative effort to draw the kind of image picture. For example, if one painter feels that he /she has more interest to draw any human face or body paint pictures as well as when he sees any one natural face or body, he can remember their body shape and face shape in his/her brain memory more clearly, even he /she also feel his/her owning human face or body image creative effort is more proficient or excellent more than to create any other kinds of things image to draw pictures. Them I may ensure that painter ought concentrate on drawing any human faces or bodies picture images in order to improve his/her painting skills more eaisly. Hencem any painter must need to know whether which aspect of picture image himself or herself mind creative effort, that he/she ought own high brain image creative
effort on this kind of thing.

I mean that when the painter discovers that he /shw own high image creative effort on matural sene, he /she ought concentrate on drawing trees, flowers, woods, etc. natural environment pictures, or he /she discovers that owning high image creative effort on furnitures, bicycles, books, cups, toys etc. different kinds of productimages.Then. this painter ought concentrate on drawing any one of these products in prder to imrpve his/her drawing painting skills. Hence, any one painter must need to know whether he/she owns which aspect of brain
image creative effort in the highest level in order to draw his/her paint pictures more easily.

Improving brain creative effort future development

Nowadays, brain scientists, brain doctors, psychologists had began to research how to improve humans ourselves brains imagine creative effort significantly. They had began to apply robots to build brains which can similar to human brains, they had been carrying experiements to research whether future artificial brains can be invented to own general human's memory ability, even imagine creative ability. In fact, they had attempted to do robot brain scientific experiments, they concluded that it is possible that future robot (artificial intelligent) brains will have chance to invent to similar human's brain to learn mind , creativity, analysis, memory ability, even future robot (AI) brain development may be improved to exceed human's general memory, mind analysis, memory, even image creative effort level. Hence, it seems that future robots can be applied to do any kinds of art creative tasks. Moreover, (AI) robot brain art imagine creative ability, it is possible that their brain art imaginative efforts can be better that human ourselves brain e.g. creative effort.

Even, future human's brain art creative efforts can not attempt to exceed robot (AI)'s brain art creative efforts forever.

I assume that future brain scientists and robot (AI) brain scientists confirm that robots may be developed their brains own high level of mind, analysis, judgement, memory and art imagine creative efforts to compare human's ourselves brains development. Then, I bring these two questions: Can future robot replace human to do any art creative tasks ? Can robots and human art creative workers cooperate in order to bring better art creative products or robots or human art creative products or robots or human art creative workers work alone can bring better art creative products, due to robot art creative workers are human art creative workers' their art creative products occupation competitors. I shall attempt to answer above these both questions as below:

If brain scientists confirm that future robots (AI) themselves brains can be invented to achieve to own high level mind, analysis, memory, art creative effort, then, they ought to used to do any kinds of art creative tasks in our societies. For example, robots can be used to apply their image creative effort to help humans painters to apply art mind to judge how to draw any kinds of paints in order to raise the paint pictures' attraction ot robots can be used to apply their music or song creative effort to help human musicians to apply art mind to judge how to write one music or song in order to let listeners can feel soft music comfortable feeling or robots can be used to apply their image creative effort to help designer to judge how to design one cloth, one house, one magazine, book, advertisement cover or any kinds of products in order to attract buyers feel the kind of product design is more beautiful to achieve purchase in preference.

I assume that future robot (AI) brain art creative effort development ought be developed to exceed human ourselves general art creative effort in possible. Although, it must be good news if robots' brains art creative efforts may be improved to invent achieve to exceed human's brain general art creative efforts, but it also brings image bad news, they may influence many art image creative workers lose their art image creative workers lose their art image creative jobs when robots can be replaced to do any kinds of art replaced to do any kinds of art creative jobs in our future societies.

Have future our art creative tasks development must need to keep balance between robot's brains and human art creative workers' brains. I suggest that human art creative workers or performances ought not feel robots may be their main art creative occupational competitors. I mean that future any kinds of art creative workers ought cooperate with robots to do any kinds of art creative tasks together. So, robots' role is future any kinds of art creative workers' assistants for one painter example, he can apply robot's art imagine creative mind to help him to mind or analyzie how to draw the natural scene picture, e.g. how to use colour, how to drawing. pens to draw the natural scene picture to be improved , it aims to let audiences can feel more attraction. Hence, the painter may apply the robot's suggestion of natural scene picture to be picture image reference. Then the painter can observe the robot's finished natural scence picture to find whether what this natural scene picture weaknesses are, in order to revise whether how to improve this robot's designing natural scene picture its weaknesses to be strengths to attact audiences' observation or raise their visal comfortable feeling or visal satisfactory feeling to this natural scene picture.

Hence, future robot's roles are only any kinds of human art creative workers' assistants. They can only been given their any kinds of art creative products to let any one art creative workers to refer or revise in order to improve their creative skills from human art creative workers. So, I feel that they ought cooperate to create any kinds of art image creative pictures together to compare that they choose to do themsleves art creative tasks alone. When they can cooperate to work together , their any kinds of art creative products must be improved to increase attraction more easily.

CHAPTER FIVE

Technology how help human to improve ability and economic development

Artificial intelligent Human clever and art creating ability methods How robots create human clever and art creating ability? Nowadays robots invention may help businesses to reduce employees number, improve performance, raise productivities, reduce cost in service industry,manufacturing industry, office , warehouse, restaurant, hotel , factory, cinema etc. different kinds of business environments, even public transport tools. However, instead of robots may bring these above advantages to any kinds of business working and service environments, whether robots may also help human to create clever and image creating ability. I shall attempt to answer this question:

On the one hand, I believe that past technology ,e.g. machine , it should not have ability to help human to create clever and image creative ability,but nowadays, robots invention that I believe it had had enough ability to help future human to raise more clever and more creating image or painting picture, art design etr. image ability, after robots had been experienced above more than ten years improvement stage from early research stage to invention stage, till to nowadays improvement stage, e.g. non-manual driving auto vehicels, even future non-manual driving skill may be improved to apply to public transport tools, e.g. trams, trains,buses, airplanes, ships etc. public transport tools, when non-manual driving skills can be improved to own the most safe driving skillful ability to compare human driving skills.

On another hand, when robots could be invented to be applied to medical or hospital surgery aspect, e.g. roboting surgerys may help surgery doctors to do complex surgery in surgery rooms, or serving patients tasks in any hospital working environments. They can help nurses and doctors to spend more time to do more important tasks urgently, so medical or surgery serving robots may help nurses and surgery doctors to reduce task load pressure and create clever or improve their surgery skills to when they can cooperate to work in hospitals.

On the other hand, robots can be invented to help any public transport drivers to avoid more traffic accidents occurrence on any countries roads. So, it seems that non-manual driving public transport tools invention may also help human drivers to improve driving skills in possible, when they can learn how to avoid sudden traffic accidents occurrence in any countries roads in any time. so, any kindsof public transport tool drivers ought learn how to avoid traffic accidents skills from future non-manual driving robots invention. Instead of non-manual driving robots and hospital patients medical care or surgery service robots may help public transport tools drivers and hospital nurses and doctors to concentrate on spending time to treat any more important and urgent matters every days. Even, future restaurants may let cooking restaurants may let cooking robots to help human cookers to cook more different kinds of good taste food, to human cookers may learn cooking robots cooking skills in order to improve themselves traditional cooking skills often, in order to compare their cooking skills between human cookers and cooking robots.

On conclusion, it seems that cooking robots ought help human cookers to create any kinds of new cooking skills. Moremove, futuer robot cookers ought be future human cookers their cooking coaches. These robot cookers will help human cookers to create clever cooking skills in possible. Also, future non-manual driving robots ought help human drivers to create new driving skills in order to improve their driving skills to reduce sudden traffic accidents occurrence easily on any countries roads in any time, future hospital surgery or patient care service robots may help surgeons or nurses to do any surgerys in surgery rooms or looking care patients in hospitals. So, when robot surgeons help human surgeons to do complex surgerys in surgerical rooms, human surgeons can learn how to do more complex surgerical tasks for every surgeons when human surgeons can observate every surgerical robots how to do surgeons together. Hence, it seems that robot surgeons also may create future human surgeons themselves innovate surgerical skills from traditional surgerical skills improvement. So, future artificial intelligent technology ought help any kinds of human occupations to create clever, even improvement themselves traditional skills to new innovative skills absolutely.

Can artificial intelligent machine help human to predict economy development

What does AI prediction machine mean? Why AI machine can make more accurate economic prediction to compare human economists? How to apply AI machine to make more accurate economic prediciton to compare human economists? In tradition, economists can gather data to make statistics to analysis how to predict this year or next year either economic growth or recession occurrence in possible.

But, since AI machine tool, it can apply to manufacturing, service , logistic , office, hospital, shopping center etc. difference business working environment aspects, whether it can help economists to make more economic prediction in possible or economists will need AI machine to help them to make more accurate economic prediction for businessmen or societies needs. I shall attempt to

give evidence to explain as
below:

Economists view our world is differently than most people. Businessmen are influenced everything by forces such as supply and demand, production and consumption, prices and costs economic concept.

However, AI machine is needed to everywhere, such as packed into your smart phones apps, warehouse logistic workers, non manual vehicles. So, AI machine may innovate our lives. Human needs AI machine to assist our tasks from complex to simple. Businessmen may bring economic time and saving long time cost benefits. Such as online shopping method, AI can help consumers find any kinds of different products prices and photos to compare in order to make reasonable purchase decision more easily from webstores channel as well as businessmen can apply ecommerce sale channel to help them to sell products to different countries in short time.

Prediction is being used for traditional tasks in business environment, such as inventional management and demand and supply forecasting. For example, because the smart phone seller predicts smart phone buyers number will reduce when he feels every family, old and young age adult who have owned at least one smart phone to every family in the country. SO, the smart phone seller ought reduce to manufacture and sell the old smart phones in order to avoid the kind of old smart phone price to reduce, due to the kind of smart phone demand purchase number will reduce,but he ouhgt to continue to innovate new kind of smart phone in order to supply new innovative smart phones products to satisfy future smart phone users needs. It is general economists judgement.Such as this smart phone case, when smart phone prediction is cheap, there has two simple economic forces drive the new opportunities, such as new kind of innovative smart phones need that predction new kind of innovative smart phones will be needed to creat in order to keep this smart phone manufacturer competitive ability in the country.

However, at low levels, a prediction machine, such as AI machine can relieve humans of predictive tasks and save on costs, when AI machine help economists or businessmen to do gather this smart phone manufacturer to manufacturing and selling this kind of smart phones data in this country consumption market every day. AI machine can gather data in short time and then it can give this kind of smart phone manufacturing and sale number data to this smart phone company every day in order to make statistic conclusion. So, AI will affect the economic of a business so dramatically, that they will no longer be used to simply enchance productivity in executing against the strategy, they will help this smart phone manufacturewr to change the strategy itself, such as whether

it ought continue to improve or innovate another kind of new smart phone design and function in order to satisfy future smart phone users needs. So, smart phones daily manufacturing and selling data is the most influential point to help this smart phone manufacturer to make decision whether it is the right time to innovate its another new kind of smart phone design and function or not in order to keep its smart phone sale competitive ability in this country.

Amazon can apply AI machine to help it reduces warehouse seeking goods time from shelves. Because Amazon has many shevles, it putted many kinds of products in shelves. Every one of worker will need long time to seek the kind of goods in order to transport to customers when they had bought the kind of goods. Amazon found this long time seeking goods problem to avoid transport time waste. It applied AI transport machine to help human worker to seek any kinds of goods from shelves. On consequently, AI transport machine help every warehouse worker to reduce at least 5 to 10 minutes to seek ths right kind of product from shelves. So, if the warehouse has 600 different kinds of products to prepare to transport to global buyers homes in the day. Every product seeking time from shelves will reduce 600 x 5 minutes or 600 x 10 minutes seeking time. So, this 600 different kinds of products will reduce 3,000 minutes or 6,000 minutes transport time to deliver to global buyers homes. So, AI machine had helped Amazon to reduce goods transport time in this day. It can help Amazon bring economic transport time shorten benefits to let global buyers can get their products within the day or tomorrow rapidly. In this case, AI machine help Amazon 's business model from shopping -then -shipping to shipping -then -shopping , AI helps Amazon to avoid buyers whose purchase goods return occurrence chance when they can be transported to their homes in short time. and accelerates the timing of investment.

Hence, it seems that AI machine may help businesses to bring these economic benefits: mahcine learning is often referred to as advances in artificial intelligence because systems prediction on this technique learn and improve over time, these systems produce significantly more accurate predictions than human economists or businessmen, traditoinal statistics methods require the articulation of hypotheses or at least of human economists analysis. In fact, AI technology is called " machine learning) for a reason. The mahines leearn form data. The prediction machine has to learn how the data is associated with actual incidences of irregular heart problem, such as Amazon transport time reducing case, when Amazon discover it will have many buyers choose either goods return decision more than non -goods return decision. AI mahcine help Amazon to seek goods from shelves to avoid goods long time seeking from human workers when AI transport delivery machine help human workers to do " goods seeking from shelves tasks working first step, then human worker only need to spend time to put the goods to lorry to transport to lorries to drive to airport in order to fly the goods to overseas customers in second time. So, AI machine may help Amazon to gather warehouse different kinds of goods shelves location data any where in order to help human workers to reduce goods seeking timr from shelves. AI mahcine may Amazon warehouse workers to avoid to spend long time to seek goods from shelves, they can concentrate on spending time to put
goods on lorries.Also, AI machine help Amazon to reduce warehouse workers number or reduce salaries expenditure, when

AI warehouse machine can replace human workers to do goods seeking tasks from shelves.

Ai machine help organizaitons to achieve new division of labour aim. Such as, a law firm may apply AI mahcine help human lawyers to take legal documents and predicted which information was condfidential. This product is valuable to law firms because when they are required to disclose documents, they have to black out comnfidential information. Historically, redaction was done by hand, with humans reading documents and blacking out confidential information. SO, AI machine help lawyers to save time and effort. Such as Amazon case, AI machine creates new division of labor strategy in any one of Amazon warehouse. AI machine will concentrate time on doing " goods seeking tasks from shelves in first step", then human workers will concentrate on time on doing " goods are putting on lorries tasks in second step ". Traditoinal , warehouse workers need to spend time to do seek goods from shelves task and putting goods to lorries tasks both. Nowadays,warehouse AI machine only need to spend time to do seek goods from shelves task , but human workers only need to spend time to do goods are putting to lorries tasks. So, AI machine will change traditional divisoin of labor to warehouse tasks in the future. Because AI machine will help workers to reduce goods transportation time and reducing goods return chance occurrence.

Printed by Libri Plureos GmbH in Hamburg,
Germany